PIT L.. _.

I would do it all over again ...

Many thanks to,

Pete Storey at the Sentinel newspaper.
Helen Burton at Keele University (William Jack Collection)
Jan Lomax, Administration.
Kevin Oakley, Resource.
Adrian and Rachel, The Choir Press.

And last but by no means least thanks to my Mum Pauline.

Brian Shufflebotham.

PIT LIFE:
I would do it all over again ...

GEORGE SHUFFLEBOTHAM

Copyright © 2018 George Shufflebotham

All rights reserved. No part of this publication may be reproduced or transmitted in any form or by any means, electronic or mechanical including photocopying, recording or any information storage or retrieval system, without prior permission in writing from the publishers.

The right of George Shufflebotham to be identified as the author of this work has been asserted by him in accordance with the Copyright, Designs and Patents Act 1988

First published in the United Kingdom in 2018 by
The Choir Press

ISBN 978-1-911589-62-4

Contents

	Acknowledgments	vi
	Preface	vii
1	Mine of Information	1
2	Fear Never Far Away from the Pit of Your Stomach	5
3	My Brush with Death in Coal Fire Explosion	9
4	Haway the Lads … Geordies Teach Us Union Ways	15
5	The "Hillbillies" who sorted out old coal faces	21
6	Ahead of its time	27
7	"Wayn Do it mar wee ternayt!" said Arthur … and we did	31
8	Lads so proud to be in pit gear	37
9	Riding a runaway underground train	41
10	I could have gone down the pit at six	45
11	Spike, Tummy and Herby … last of the funny kind	51
12	Pit Life? I would do it all over again …	55
13	Passage to Florence	61

My own feelings, and I hope those of many of my old colleagues, are summed up in this poem by my wife Pauline which she has entitled

They Can't Take Away The Memories

I stood and watched with heavy heart
At the pit winding wheel just out of town,
I thought of the lives that were torn apart
On that sad day they closed them down.

Rough times, tough times, but sometimes good
Are my memories of those days below,
Bawling and shouting, wanting blood!
You have to have been there just to know.

But the sweat and toil's now gone forever,
The sweltering heat, the dirt and grime,
But the bonds of workmates none can sever
They will remain till the end of time.

Preface

My Dad George Shufflebotham spent thirty six years in the pit. He worked at five North Staffordshire Collieries, rising from a shot firer at Berry Hill Colliery to Senior Overman at Florence Colliery.

In this series of articles written between 1996 and 2003 Dad recalls his experiences underground and reflects on the many human aspects of a working life down the pit. Coal has been won by the indefatigable miners of North Staffordshire for over seven hundred years. The men in these stories were the last of a mining brotherhood. This body of work pays tribute to them and the communities from which they came. And I am proud to be a miner's son.

These popular articles were featured in the Sentinel newspaper in the series "The Way We Were" and "All Your Yesterdays" presented by John Abberley.

Brian Shufflebotham

Davy Lamp no. 4 carried by George Shufflebotham.
Courtesy Brian Shufflebotham

1
Mine of Information

Like so many other lads in North Staffordshire, I followed my father down the pit, so I knew what to expect when I rolled up for my first day at Berry Hill Colliery.

My father had died when he was 42, so as a family we were pretty hard up. To be honest, I went into the pit mainly for the big money.

After changing into my pit clothes, I got my lamp and checks and waited in the queue for the cage to go down below. We were crammed in like sardines and the cage dropped down the shaft like a house brick. My heart came up into my mouth and my ears popped. Halfway down we passed another cage going up and felt the pressure as the two cages passed each other at such a speed.

The senior overman took me to my first job at the top of Carr's Cut, a one-in-five dip. I was to be the lasher-on and lasher-off. I had to lash up the tubs as they came up the dip on an endless rope, which disappeared down a black abyss.

While doing this job I met an old chap named Bill Pepper who'd worked at Berry Hill for many years. He introduced me to a tin of snuff. It nearly blew my head off. Then I was initiated in chewing twist.

My next job was called "behind the pit" which involved coupling up empty tubs at the pit bottom. I worked in a little tunnel about 5ft 6in by 5ft wide. In this confined space I worked non-stop. It was a dark and lonely job.

PIT LIFE: *I would do it all over again ...*

Looking like a bleak mountain, the tip at Berry Hill 1965
Courtesy of *The Sentinel*.

After a few weeks I'd had enough of it. I applied to go on the coal face. The training officer said I was a big lad and he'd put me at the front of the queue. I was put on a face called 5's Moss.

I was put alongside Jess Whitehurst, a real old collier from Haden Colliery, Cheadle. It was a pick and shovel job, all hand-loading. The coal was shot-fired the previous night, but there was no machinery.

I was 18 and like a raging bull to get at the coal. Jess said "Ay lad, dunner get carried awe. Let's have some timber up. I'm goin' to learn you right."

It was like the league of nations on the face. There were Italians, Germans, Poles and a chap named Dai from the Rhondda. The Poles were fantastic workers. It was probably something to do with their harsh background. Also, I can't speak too highly of the Italians.

You had to watch out for mice if you hung your food and donkey coat on the roadway. You often found that your butties had gone, with just the shape of them left in the paper. And you had to shake your donkey coat because the arms and pockets were always full of mice.

In the 1950s accidents were commonplace. One that sticks in my mind was a packer being buried by a roof fall. There was a lump of dirt on him as big as a car. All we could see was his feet.

We expected the worst, but fortunately the ground was a bit soft and a steel bar and post had fallen across him, protecting his body. But it was still a miracle that we got him out alive.

On many occasions I travelled underground from Berry Hill up to the old Top Pit, close to the area of Eaton Park estate. An underground river ran down the side of the old roadway. For us, it was a second means of egress, or getting out.

Old Top Pit was made of wood and lagging and the timbers were full of grubs, millions of them. It made me shudder when we came up that way in the cage.

In my father's day, Top Pit had a shaft which was like a corkscrew. The men actually walked into the pit on a circular path which went down to the bottom of the shaft.

Going back to my first experiences on the face at Berry Hill, we came up the pit completely knackered. We had a wash and shower, with three or four of us sharing one shower.

Then we went to the canteen, where you get a proper hot dinner for ten pence or a shilling. But most of us just had a homemade meat and potato pie in a small dish filled with gravy.

PIT LIFE: *I would do it all over again ...*

At the end of the counter there was a great jug of gravy and about six loaves of cut bread. You could have as much of this as you liked for dip. It was absolutely delicious.

2

Fear Never Far Away from the Pit of Your Stomach

Besides being hard work with a pick and shovel, it was also extremely hot on the face at Berry Hill. We drank a lot of water and worked practically naked.

On the day shift we took seven pints of water down the pit and by half past nine or ten o'clock it had gone.

More water was brought to the face to fill up the banjo bottles. These were round bottles of different sizes we kept slung round our necks.

Berry Hill Colliery.
William Jack Collection

PIT LIFE: *I would do it all over again ...*

But first, we had to get to the face. We walked from the cage at the pit bottom to the riding dips, which were often very low and in bad condition. They were what we call return airways.

We had to lie down flat on little trolleys, with the roof just inches above our heads. If you think the Black Hole of Calcutta was awesome, you should have tried riding on one of those trolleys.

You travelled like this for about half-a-mile, perhaps more, and the trolleys didn't half move. If you had pushed your head up more than four or five inches you'd have had it knocked off.

I've crawled into these return levels when they were totally disintegrating. You'd be going through a hole 18 inches high by two feet wide. And the ventilation was like a storm.

Just a couple of inches above your back was solid steel and dirt. If you'd thought about claustrophobia, you'd have gone mad. If you'd panicked and tried to stand up there was no way out. There were umpteen men in front of you and behind you.

So you had to keep cool going through these narrow roadways. It still makes me shudder to think about it even though it was 40 years ago.

Then there was a final walk of a mile or so to the coal face, so eventually you'd be two or three miles from where you started.

The first job was to make a passage through the coal, so there was a way in and way out. The coal had already been blasted on the night shift.

So we had a pick and shovel, a hammer, a lamp hanging between the legs, a banjo bottle round the neck and snap-box on the belt, half-crawling and scrawling to the face. It was murder getting onto your buttock, or section. And you were on your own.

The face would be 150-200 yards long and when we started

Never Far Away from the Pit of Your Stomach.
Courtesy of *The Sentinel*

work there was only about two-foot headroom. After half-an-hour, when we'd broken through the coal, we had a height of about four foot six, which was much better.

Because it was so hot, we were usually naked when doing this job. You couldn't see in front of you for dust. There was just the narrow beam from your lamp. And there was a constant noise from the conveyors.

I can't remember seeing many fat miners. They were like Arabian donkeys, all ribs.

Clothes only lasted a few days in the pit and neighbours used to drop off old trousers, shirts, odd socks, anything. If a bloke came to the pit with a new pair of boots, we'd see if one of his old ones could make up a pair of boots, or clogs, for somebody else.

I've known men go down the pit with a clog on one foot and

a boot on the other. In fact, there's many a time when I've worked in odd boots, odd socks, odd anything!

There was a great character at Berry Hill called Harold Heath. He'd had a serious accident in a roof fall and was paralysed below the waist. But he was a hard man and fought his way back. He was given a job on the surface.

They thought Harold would never get out of his wheelchair, but he was very strong-willed. He walked with leg irons and even learned to ride a bike. All us young miners looked up to him. If you wanted any advice you went to Harold.

On Christmas Eve, the last day of work before the holiday, there was usually a low turnout, but the festive spirit was there all right, even down the pit.

At Berry Hill we'd find that the night shift had hung Christmas decorations in the pit. This was in contravention of safety rules, but the management turned a blind eye.

There was no booze down the pit, of course, just mince pies and cakes. We used to meet for a booze-up in the pubs near Berry Hill. These nearly always ended with somebody playing the piano. It was a right fiasco but very enjoyable.

What comradeship we had in those days!

3

My Brush with Death in Coal Fire Explosion

In 1960, rumours were hot that Berry Hill Pit was going to close. I found a note on my lamp to see Tom Smith, the training officer. He told me I was being moved to Mossfield Colliery, Longton – or Old Sal.

A chap told me the name of Old Sal originated in the last century. It was either the nickname of a beam engine which pumped water out of the pit, or of a woman named Sarah who watched over the engine for her husband while she did her knitting at the pit top.

Another theory is that the name might have derived from Salt's Cottage close to Mossfield Colliery because the French for salt is sal.

Like Berry Hill, Mossfield was an old-fashioned pit, with manriding trolleys, low roofs, and a lot of dips. Everything worked on tubs and haulage. It did have one steel-plate belt, which broke on numerous occasions and came tumbling down the dip like a sheet of fire.

I was 20 and a fully-fledged collier, but at Mossfield I had to go odding about the pit. To get a regular job you had to see the senior overman George Foster, a smallish chap who ruled with an iron hand.

After I'd been mithering George for six months he finally gave me the job I wanted – loading coal, on the new Holly

Lane face. I worked with a chap named Stan Holford. One day I had a brush with death.

We'd more or less drawn off and had about four yards of coal left. It was hard and hadn't fired very well, so we drilled three holes in the coal to fire it again.

A shotfirer charged up the holes with powder and coupled two of them up to fire together, leaving the third to be fired on its own. We retreated to a safe position. After firing the shots he flashed his light and we went back to the face.

I took a pick to the coal and still thought it hadn't fired very well. Then I noticed some wires still coupled to one of the shots. I tried to grab the wires. As my hand went down the shot went off.

I was blown backwards down the face. It was like going down a time tunnel. I have never seen so many coloured lights. It knocked me unconscious, but I came round after being carried into the dip, where the intake air was cooler.

We had been working with hardly any clothes on and I was covered all over in cuts. I still have a few scars to this day. But

My brush with death in coal face explosion at Mossfield Colliery.
William Jack Collection

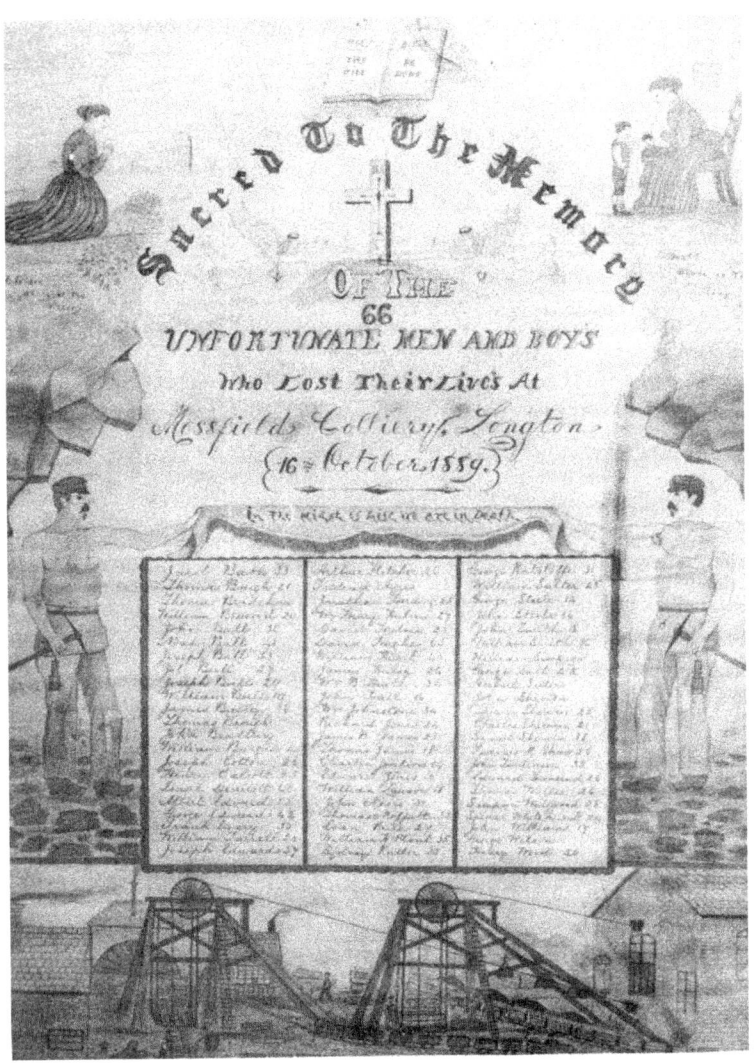

An ornate pencil memorial drawn by Victorian artist George Woodward in 1890, commemorating the sixty-six men and boys who died in the October 16th, 1889 disaster at Mossfield Colliery.

when I was questioned about it, I denied that the shot-firer's action had caused my wounds because he would have been sacked.

Mossfield, of course had a history of underground accidents. The worst was the explosion in 1889 which killed 66 men and boys. There was another in 1940 in which 11 men died. The pit bottom fireman showed me the places where these explosions occurred. We couldn't get in there because they were bricked up. I've thought a lot about those places and the atrocities of the deaths of all of those men.

The fireman also took me down an old roadway to Stirrup and Pye's Colliery, one of Mossfield's means of egress. It was amazing. You went through a tunnel dug out of solid rock, with no supports.

Explosions occur in the pit when coal dust is ignited. Coal dust, of course, also caused lung disease. And stone dust, which comes off ripping or packing, is even more damaging. It tears at your lungs.

I knew men in their 60s on the coal face who could hardly breathe after inhaling all that coal dust over the years. So I listened to a chap who told me about the benefits of chewing tobacco. If you chew tobacco, he told me, you'll lose your teeth. But if you don't chew it, you'll lose your lungs because you'll breathe through your mouth.

When chewing your baccy, you breathe through your nose. So as you are continually blowing it, the dust is filtered. Coupled with a bit of snuff, it keeps your nose clear.

However, the baccy doesn't do your teeth much good. On top of that, I started suffering from heartburn. It was costing me a fortune in Rennie's.

The old collier told me the cure for this was to suck a lump of coal. It worked.

The Way We Were

Haway the lads... Geordies teach us union ways

WHILE working at Mossfield Colliery in the early 1960s, GEORGE SHUFFLEBOTHAM had his first encounter with Geordie newcomers. In his latest article he recalls the impact made by miners from the North East, tells of a ghostly apparition underground and remembers the vain attempt of a fellow pitman to stop torture by dripping water at the coal face.

WHILE I was having a shower after finishing the noon shift, I heard an accent in the pit baths I'd never heard before. The Geordies had arrived. It was around 1962.

[text continues, largely illegible]

Pick

[illegible column text]

Faded

[illegible column text]

Cover

[illegible column text]

WHEN there were no more faces at Mossfield, we knew the writing was on the wall. Then they drove a roadway from Mossfield to the backyards of Shaftesbury Colliery and we knew it was going to be finished.

None of us wanted to go to Holly Bank anyway, where the next jobs were. So the chargehands of the Bashleyet seam and the Holly Lane seam were called together to discuss it.

The Bashleyet chargehand for the last six or so had to leave Mossfield for The Holly Lane then he moved on to Hem Heath a few months later.

More Mining Memories next month

* Above: The Durham Gala is shown on this NUM banner.

* Below: Working in a cramped seam a few feet high ... of the type the Geordie miners had been used to.

Pictures from The Miners by Anthony Burton (Future Publications Limited)

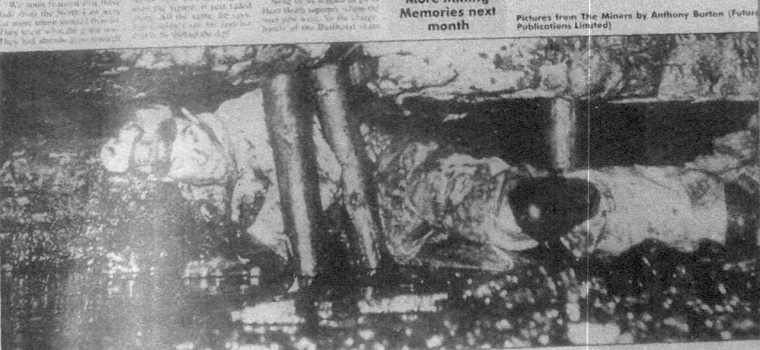

4

Haway the Lads ... Geordies Teach Us Union Ways

While I was having a shower after finishing the afternoon shift, I heard an accent in the pit baths I'd never heard before. The Geordies had arrived. It was around 1962.

From the start, these lads mixed in very well with us, being the same kind of people as ourselves. They certainly added some colour to life in the pits and to the life of the city generally.

Lying down.
Picture courtesy *The Miners* by Anthony Burton (Futura Publications Limited)

PIT LIFE: *I would do it all over again ...*

The Durham Gala is shown on this NUM banner.
Picture courtesy *The Miners* by Anthony Burton (Futura Publications Limited)

Leaving their families at home in many cases, the Geordies stayed in bed and breakfast places all over the Potteries until they found themselves a new home. This touched the hearts of North Staffs miners.

The newcomers soon showed us they could work, too, although they found it difficult to adjust to working on coal seams six feet thick. Up in Durham they had been used to working on their bellies. Here they had to use a pick and shovel standing up.

They also showed us how to celebrate Hogmanay. And they introduced us to their delicacy, peas pudding, which is ham and peas done in a pie. They used to bring it down the pit, I had many a slice.

I know they respected us for being hard workers, although I remember one Geordie making the remark "Youse Staffies may be good workers, but youse'll work for anything!" meaning that we were a soft touch.

We soon realised that these lads from the North-East were far more union-minded than us. They knew what the game was. They had already gone through the problems of pit closures at home in Durham.

If there was any negotiating to be done with management they were the best ones to do it. They did a tremendous amount to get improved conditions and pay and made us locals more aware of our union rights.

In those days we were all promised a job for life and lads still followed their fathers down the pit. Good teams were head-hunted by colliery managers, who offered them more money. So men moved about a lot.

I knew our time at Mossfield was limited because it was a small uneconomical pit.

Mossfield Colliery.
William Jack Collection

PIT LIFE: *I would do it all over again ...*

In the short time I was there I heard a genuine story about a ghost. The man swears he saw it and even now, years later, still says he'll never forget it.

A team of electricians went into the pit to do a job which required the power to be switched off. To do this, a man went down on his own to an old part of the mine workings.

As he knelt down in front of the panels to switch everything off, he felt a presence, though no-one was in the area. Moving his head slightly, he saw a figure of a man out of the corner of his eye.

The apparition was only a few feet away, but this chap kept his head down and he got on with the job for a few minutes. He then had to go past the figure to get out again.

The electrician plucked up courage and started to move. As the light from his lamp fell onto the figure, it just faded away. All the same, he says you couldn't see his feet for dust as he shot up the dip!

From what he could gather, some years earlier an old deputy went down this roadway regularly to pump water out of the bottom of the dip. One day there must have been a quantity of methane there because the man was overcome and died.

There was plenty of water about on Mossfield. When I worked on the Bullhurst seam, on what we called the deep side, it was always flooded and we worked loading coal standing in a foot of water.

We stripped off for this job and got stuck in. We shovelled half water and half coal. The water drained down the coal face and dripped out of the roof. As it contained chemicals it had a stinging effect on the skin.

This water torture was getting one of my mates down. He decided to cover us up. He couldn't find anything else but a large piece of mesh, which he put over our heads as we shovelled.

Naturally, it did absolutely nothing to stop the flow of water, which kept running down on us, through the mesh. It was a long time before my mate lived that down!

When there were no more face developments at Mossfield, we knew the writing was on the wall. Then they drove a roadway from Mossfield to the working of Florence Colliery and we knew it was going to be flooded.

None of us wanted to go to Hem Heath superpit, where the next jobs were. So the charge-hands of the Bullhurst seam and the Holly Lane seam were called together to toss a coin.

The Bullhurst chargehand lost the toss, so we had to leave Mossfield first. The Holly Lane men followed us to Hem Heath a few months later.

Hem Heath and Florence pit (Colliery).
Courtesy of *The Sentinel*

The Way We Were

GEORGE SHUFFLEBOTHAM found himself in a different world when he was transferred from Mossfield Colliery to Hem Heath. In this article of a series, the much-travelled former colliery official recalls an accident involving a workmate which turned into a race and his own close brush with death in a roof fall.

The 'hillbillies' who sorted out old coal faces

WHEN I moved to Hem Heath in 1963, everything about the "superpit" looked enormous compared to anywhere else.

It isn't a personal family pit like Hill at Mossfield, where you knew friends. There were miners from over the country and surgeons too. It made you feel you were nobody.

On my first day there after being transferred from Mossfield, a lot of the men our way about finding the pit. It must have been obvious I'd never seen anything like it.

I got halfway in it was like step into a ballroom, with the electric and a locomotive waiting drawing cars. At Mossfield, we on our hands and knees getting out of the cage.

After we'd travelled a couple of underground we soon learned though everything was so modern surface, there were old faces worked at Hem Heath which far in the past.

We were stopped and we were it out, we found a five-foot which took us back to the part from the coal cutter. I was done by hand, it was a hard stood job.

As we knocked this old face after a few months and moves to the Hem Heath.

D on this face for a couple and halfway through the one day there was a serious tough as things turned out it like a cloudy of cries.

I came up that a man had been hit by a roof fall. We got the man out, dressed his wounds and find him on a stretcher. Then everyone stood back a moment before lifting him.

I should explain that because of the distance covered and the heat it was the practice in these circumstances to have 10 men acting as stretcher bearers, chosen from those who had finished their tasks.

There was no shortage of volunteers, as it meant the men chosen would get out of the pit for an early start. And with this in mind, one old collier named Tom suddenly came running over a pile of dirt.

He didn't see the stretcher and planted one heavy boot on the injured man's chest and the other boot in his groin, making the casualty groan loudly.

Then, as Tom carried the rear of the stretcher, the partly-filled five-pint bottle tossed his neck started to swing, striking the casualty on the back of the neck. To add insult to injury, the bearers took their seals off and piled them up on top of the injured man.

The man might have had a narrow escape in the roof fall, but his mates nearly killed him!

ON the latter face, with the conditions deteriorating, I came near to meeting my Maker. It was my turn to go on what was known as the wooring up, a small road parallel with the main face which takes the weight off the main roadway.

As I started to clear a site, there was a whoof! and the coal face and roof collapsed, with me underneath it. There were none of the warnings like creaking props, or bits of dirt falling off the roof. A shower of coal and stone came down, pressing me into the floor.

Luckily for me, there was slack and small coal on the floor of the seam. But I couldn't blink and could hardly breathe. I was drifting in and out of consciousness. All my family life went through my mind, like a video tape on a fast rewind.

I couldn't move a muscle. I seemed to be under the coal face for a lifetime. But in fact it was only an hour or two before I heard my mates digging and shouting. I'm sure that I owe my life to those men.

But my mother, who is nearly 90,

● Hem Heath Colliery, which opened in 1924

has always maintained that I had an angel on my shoulder that day, my grandmother.

AT Hem Heath, I had my first experience of a dispute. It was normal for a man to earn 65 shillings in 70 shillings a shift for packing. However, myself and two mates were working in competition with each other and doing well, earning £5 to £6 a shift.

One day we turned up for work to find that hundreds of men were refusing to go down the pit because of a dispute over the rates for packing. We there kept a low profile at the bar with our heads bowed.

Not long after my accident, the colliery manager asked me if I would interested in training at a colliery of coal. So I went to college at Stoke for six months and became a shotfirer.

I was about 23 at the time and had to choose between being one of the boys or being in charge. I knew I had to move on.

When the chance came, I decided to go to Wolstanton.

● Photographs from Mining Memories by Fred Leigh (S. Publications)

with machinery

> 'After we'd travelled a couple of miles underground we soon learned that although everything was modern on the surface, there were old faces being worked at Hem Heath which belonged far in the past'

More Mining Memories next month

5

The "Hillbillies" who sorted out old coal faces

When I moved to Hem Heath in 1963, everything about the "superpit" looked enormous compared with anywhere else.

It wasn't a personal family pit like Berry Hill or Mossfield, where you made close friends. There were miners from all over the country and umpteen coal faces. It made you feel you were just a number.

On our first day there after being transferred from Mossfield, a lot of flak came our way about being the hillbillies. It must have been obvious that we'd never seen anything like it before.

At the pit bottom it was like stepping out into a ballroom, with the electric lighting and a locomotive waiting with manriding cars. At Mossfield, we'd been on our hands and knees after getting out of the cage.

But after we'd travelled a couple of miles underground we soon learned that although everything was modern on the surface, there were old faces being worked at Hem Heath which belonged far in the past.

When the loco stopped and we were told to get out, we found a five-foot coal face which took us back to the 1940's. Apart from the coal cutter, everything was done by hand. It was a real pick and shovel job.

PIT LIFE: *I would do it all over again ...*

Coal cutting of yesteryear.
Photograph courtesy of *Mining Memories* by Fred Leigh (SB Publications)

However, we knocked this old face into shape after a few months and gained the respect of the Hem Heath men.

I stayed on this face for a couple of years and halfway through the noon shift one day there was a serious accident, though as things turned out it was more like a comedy of errors.

The word came up that a man had been hit in a roof fall. We got the man out, dressed his wounds and tied him on a stretcher. Then everyone stood back a moment before lifting him.

I should explain that because of the distance covered and the heat it was the practice in these circumstances to have 10 men acting as stretcher bearers, chosen from those who had finished their tasks.

The "Hillbillies" who sorted out old coal faces

There was no shortage of volunteers, as it meant the men chosen would get out of the pit for an early pint. And with this in mind, one old collier named Tom suddenly came running over a pile of dirt.

He didn't see the stretcher and planted one heavy boot on the injured man's chest and the other boot in his groin, making the casualty groan loudly.

Then, as Tom carried the rear of the stretcher, the partly-filled five-pint bottle round his neck started to swing, striking the casualty on the back of the neck. To add insult to injury, the bearers took their coats off and piled them on top of the injured man. The man might have had a narrow escape in the roof fall, but his mates nearly killed him!

On the same face, with the conditions deteriorating. I came near to meeting my Maker. It was my turn to go on what was known as the scouring rip, a small road parallel with the main face which takes the weight off the main roadway.

As I started to clear a site, there was a woof! and the coal face and roof collapsed, with me underneath it. There were none of the warnings like creaking posts, or bits of dirt falling off the roof. A shower of coal and stone came down, pressing me into the floor.

Luckily for me, there was slack and small coal on the floor of the seam. But I couldn't blink and could hardly breathe. I was drifting in and out of consciousness. All my family life went through my mind, like a video tape on a fast rewind.

I couldn't move a muscle. I seemed to be under the coal for a lifetime. But in fact it was only an hour or two before I heard my mates digging and shouting. I'm sure that I owe my life to those men.

But my mother, who is nearly 90, has always maintained that I had an angel on my shoulder that day, my grandmother.

PIT LIFE: *I would do it all over again ...*

Hem Heath.
Courtesy of William Jack Collection

The "Hillbillies" who sorted out old coal faces

At Hem Heath, I had my first experience of a dispute. It was normal for a man to earn 65 shillings to 70 shillings for a shift for packing. However, myself and two mates were working in competition with each other and doing well, earning £5 to £6 a shift.

One day we turned up for work to find that hundreds of men were refusing to go down the pit because of a dispute over the rates for packing. We three kept a low profile at the back with our heads bowed!

Not long after my accident, the colliery manager asked me if I would be interested in training as a colliery official. So I went to college at Stoke for six months and became a shotfirer.

I was about 23 at the time and had to choose between being one of the boys or being in charge. I knew I had to move on.

When the chance came, I decided to go to Wolstanton.

The Way We Were

TEAMWORK saves the life of a miner who was seconds from death ... a ghostly figure appears in an old roadway ... and a lost starting-handle to a new cutting machinery is found in a burning brazier.

These are among the latest tales of working life underground from former colliery official GEORGE SHUFFLEBOTHAM, who moved to his fourth pit, Wolstanton, in the 1960s.

Ahead of its time...

●Wolstanton Colliery in the late 1950s during work to modernise

WHEN I arrived at Wolstanton Colliery, modernisation had just been completed. Besides being the deepest pit in Britain it was huge, with roadways stretching miles to Hanley Big Pit, Sneyd and beyond.

The colliery manager at the Sneyd pit was John Bacharach, a forward-looking young man. As a face deputy, I was put on a face just coming into production in Hard Mine seam. It was way ahead of its time.

As we travelled to Sneyd we passed Hanley and Sneyd men walking towards us. There was a massive workforce. We were miles from the pit bottom at Wolstanton.

One unusual feature I remember on the record level from Sneyd was some apparatus on a very steep incline which looked like something on a ski slope.

To help us to walk up the dip, which was about one in three or four, there was a continuous running rope. Stacked at the bottom were loads of walking-sticks, which we clipped on the rope to be pulled up.

ON another face we added to the modern technology by installing a nuclear-armed coal-cutting theatre costing mega-bucks, accompanied by scientists from the mining research establishment.

It was a Sunday dinner time and the job was practically completed, ready for shearing on the Sunday night shift. But there was one thing missing — the starting-handle.

This was like a very large wheel winch winder, some three feet long with gears down on them. It was the only one of its kind in the country. So panic set in when we couldn't find it.

After an hour's search with all men we still hadn't found it. Practically, I went up to the surface and ran out of the pit. It was a freezing cold winter's day with snow on the ground.

I decided to go back down the pit, but while waiting for the cage I spoke to the banksman, who wanted to know what was going on. I told him we had lost the starting-handle.

"Duzt 'eat," he said. "I've sayn summat like that. Duze maya mat pokes?" I followed him to a large brazier, which was burning. Standing by it was the starting-handle. The banksman had been using it to poke his fire.

A good wire-brushing and a bit of oil and it was OK. We could start up the expensive shearing machine.

LIKE most pits, Wolstanton had a ghost story. I never saw it myself, but a joint driver told me about the time he eyed off with empties at the end of the noon shift.

As he approached an old roadway there was no-one else in that part of the pit. Suddenly, a man was standing in the middle of the track. The driver slammed in his brakes, but with 25-30 mine cars behind him he couldn't stop.

The loco hit the figure, flattening him. T...
looked ...
nothing ...
However ...
had two ...
years ea...

FINA...
you ...
which ...
after h...
powerf...
inch se...
As n...
ing the ...
rang. T...
on the f...
way, ne...
I met c...
they we...
"I coo...
er. He g...
cal shor...
had gon...
The hol...
flop the...

With ...
conscio...
with the...
tion. For...
We wer...
four dips...

The b...
the spot...
as a sho...
tion and...
surface ...
medics ...

At th...
ferred so...
an emot...
put throu...
was a ve...

After...
he had g...
body. If ...
pit only...
he would...

I was ...
Wolstan...

'Sud
stan
o
driv
brak
min
he

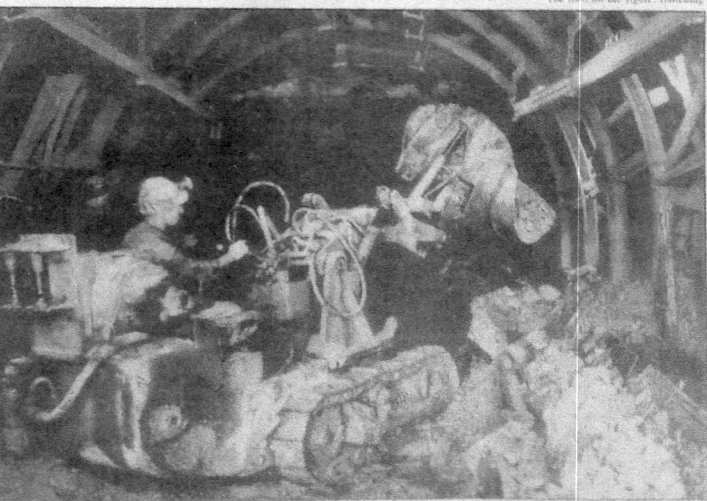

miner operates an "Eimco 623H high discharge side tipping" loader.   Picture by the National Coal Board, Scottish Division

6
Ahead of its time

When I arrived at Wolstanton Colliery, modernisation had just been completed. Besides being the deepest pit in Britain it was huge, with roadways stretching out miles to Hanley Big Pit, Sneyd and beyond.

The colliery manager at the Sneyd end was John Bacharach, a forward looking young man. As a face deputy, I was put on a face just coming into production in Hard Mine seam. It was way ahead of its time.

As we travelled to Sneyd we passed Hanley and Sneyd men walking towards us. There was a massive workforce. We were miles from the pit bottom at Wolstanton. One unusual feature I remember on the return level from Sneyd was some apparatus on a very steep incline which looked like something on a ski slope.

To help us to walk up the dip, which was about one in three or four, there was a continuous running rope. Stacked at the bottom were loads of walking sticks, which we clipped on the rope to be pulled up.

On another face we added to the modern technology by installing a nuclear-armed coal-cutting shearer costing megabucks, accompanied by scientists from the mining research establishment.

It was a Sunday dinner time and the job was practically completed, ready for shearing on the Sunday night shift. But there was one thing missing – the starting-handle!

This was like a very large wrist watch winder, some three

PIT LIFE: *I would do it all over again ...*

Ahead of its time – Wolstanton.
Courtesy of *The Sentinel*

feet long with gears down its stem. It was the only one of its kind in the country. So panic set in when we couldn't find it.

After an hour's search with all men we still hadn't found it. Frantically, I went up to the surface and ran out of the pit. It was a freezing cold winter's day with snow on the ground.

I decided to go back down the pit but while waiting for the cage I spoke to the banksman, who wanted to know what was going on. I told him we had lost the starting-handle.

"Dust 'ear," he said. "I've sayn summat like that. Dust mayn mar poker?" I followed him to a large brazier, which was burning. Standing by it was the starting-handle. The banksman had been using it to poke his fire.

A good wire-brushing and a bit of oil and it was OK. We could start up our expensive shearing machine!

Like most pits, Wolstanton had a ghost story. I never saw it myself, but a loco driver told me about the time he sped off with empties at the end of the noon shift.

As he approached an old roadway there was no-one else in that part of the pit. Suddenly, a man was standing in the middle of the track. The driver slammed on his brakes, but with 25-30 mine cars behind him he couldn't stop.

Ahead of its time

The loco hit the figure, flattening him. The driver got out shaking and looked under the train, but there was nothing. Nobody was every found. However, it transpired that someone had been killed at that exact spot a few years earlier.

Finally, this time, I want to tell you about a piece of teamwork which undoubtedly saved a man's life after he'd been badly injured by a powerful loading shovel with seven inch teeth on the bucket.

As noon shift overman, I was leaving the pit bottom when the telephone rang. There'd been a serious accident on the Bowling Alley face. I ran all the way, just under a mile, telling the men I met coming off the shift to stay where they were and make up teams of four.

I found the injured man on a stretcher. He'd been trapped by his mechanical shovel and one of its large teeth had gone straight through his stomach. The hole was so big that we couldn't stop the bleeding.

With the man drifting in and out of consciousness, we started a relay run with the stretcher back to the pit bottom. Twenty to thirty men were involved. We went flat out, even up the one in four dips.

The blood was literally running off the stretcher and the man was as white as a sheet. We got him to the pit bottom and the cage shot him up to the surface, where a doctor and paramedics were waiting.

At the pithead the man was transferred straight into an ambulance and an emergency operation was carried out there and then. He survived, but it was a very close thing indeed.

Afterwards, the paramedics told me he had practically no blood left in his body. If he had been brought out of the pit only a few seconds later, they said, he would have been dead.

I was very proud of those miners at Wolstanton that day.

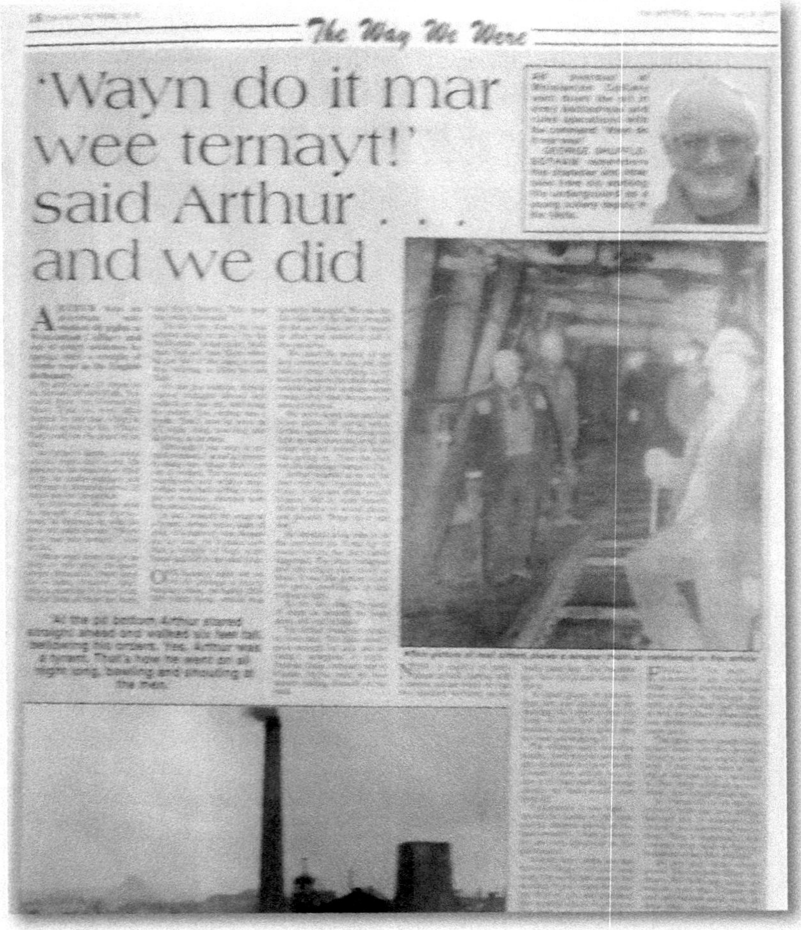

The Way We Were

'Wayn do it mar wee ternayt!' said Arthur... and we did

"At the pit bottom Arthur stared straight ahead and walked six feet bit, bellowing his orders. Yes, Arthur was different. That's how he sent us all night long, bawling and shouting at the men.

7

"Wayn do it mar wee ternayt!" said Arthur ... and we did

Arthur was an overman who worked all nights at Wolstanton Colliery and out of every sentence he spoke, only a couple of words were in the English dictionary.

He arrived at 10.10pm on the dot and left at 6.45am. You could have set Greenwich Mean Time by him. Men stepped to one side when he walked across to the offices. They could see the anger in his face.

As Arthur's deputy I sorted out the night shift work. He glanced at the work book filled in by the under-manager and then sent it skimming about 30 yards into the lamphouse.

As if it wasn't obvious what he thought of these instructions, he followed up with his own pet catch-phrase: "Wayn do it mar wee ternayt!" And we did.

Arthur went down the pit an hour or two after the men, always dressed in 1940's army kit – tunic, trousers, shirt, even a great coat if it was cold. He got them all from the Army and Navy Stores. This man was ready for battle!

On his way down he was offered mugs of tea – by the banksman, lamp man, bath man, first aid man. Even when he got out of the cage, a man was waiting to offer his tea flask.

At the pit bottom Arthur stared straight ahead and walked six feet tall, bellowing his orders. Yes, Arthur was a tyrant. That's how he went on all night long, bawling and shouting at the men.

PIT LIFE: *I would do it all over again ...*

Wolstanton 1968.
Courtesy of *The Sentinel*

Although I was only in my mid-twenties, I acted as referee to make sure things didn't get out of control. Yet even after a murderous row with a man, Arthur was back within a couple of minutes, chatting with him about football.

In fact, beneath his image as a tyrant, Arthur had a heart of gold. If somebody was desperate for coal, he'd make sure that a couple of bags were found and left in the man's car.

One Sunday night we set out to change a heavy haulage chain, probably 250-300 yards long, which was severely damaged. We ran this old chain off the face, brought up the new chain on its wooden drum and started to pull it into position.

We used the power of the face conveyor for this job, but had to keep switching it off and on because the chain easily twisted

"Wayn do it mar wee ternayt!" said Arthur ... and we did

and the wooden drum swung every time the conveyor moved forward.

We were doing fine and had done about 60 yards when Arthur appeared, flashing his light up and down the level. He came up and asked us what was going on. Then the dust and the language started to fly.

Coalcutting of yesteryear. This picture shows a scraper chain.
Courtesy of *Mining Memories* by Fred Leigh (SB Publications)

PIT LIFE: *I would do it all over again ...*

Arthur ordered us to let the conveyor run continuously. I tried to explain what would happen. But he went beserk, threw pieces of wood about and shouted "Wayn do it mar wee!".

He shrieked at the men to let the conveyor go. It ran for 10 yards before the inevitable happened. The drum collapsed and the chain tied itself in knots. It was like getting a cotton spool, ravelling it up and pulling it tight.

To this day, that 250 yards of chain lie beneath the Asda store, still tied in knots.

Yet Arthur found his excuses even then. The chain wasn't good enough, he said, and we hadn't wrapped it right. Parsons chain company was to blame. In the end, he was almost patting himself on the back.

Now, a couple of tales about myself, starting with a narrow escape when we had two shearers working on the Holly Lane face. One machine got fast solid and wouldn't move.

A small piece of granite-type dirt was sticking in the tracking. As I tried to free it a lump of dirt fell onto the machine, pushing it down onto my hand and trapping it.

The shearer had a direction handle, clockwise to move up, anti-clockwise to move down. I wasn't sure which way to turn it. If I had made the wrong choice, my hand would have been off.

Fortunately, I was right. I moved the handle to the right and the machine moved off my hand instantly. I went to hospital and was operated on for dislocations.

In those days, when you had busted fingers or cuts and scars, the worst part wasn't the accident. It was when the sister got hold of you and scrubbed the wound clean to prevent blue scars.

She got the injured part under the tap and used a scrubbing brush to get all the black out. It was murder. She scrubbed till it was pink, or you had a scar for life.

"Wayn do it mar wee ternayt!" said Arthur ... and we did

Then she stitched it up with a needle and cotton, without any freezing or injection.

Finally, for now, I remember an occasion when a face conveyor broke down halfway through the shift. A drive shaft had broken in two, but I knew where there was another one a few hundred yards away.

The spare part I needed was under some wooden boards along a very hot return roadway. I told some men at the top of the face to go and fetch it. But they refused point blank. And I could guess why.

So, I went off to do the job myself, with the men tailing behind me. I reached the boards and started to move them. Then it happened. I was covered from head to foot by hundreds of bat-like crickets.

As my cap lamp shone down, the crickets filled the beam. But with my collar up and my eyes shut, I found the part and backed away, with all these black creatures flying up and down the roadway.

I could hear a lot of laughing going on behind me.

The Way We Were

It used to be a tradition for Potteries lads to follow their fathers down the pit. But in the 1950s school-leavers signed up as trainee cleanminers because it was a well-paid job with prospects. Retired mining official GEORGE SHUFFLEBOTHAM looks back at his own experiences as a raw recruit at the Kemball Training Centre at Fenton.

Lads so proud to be in pit gear

AS SOON as we arrived at Kemball for training, we were given the basics — pit hat, pit boots with steel toecaps and a pair of bib and brace overalls.

We were really proud of wearing the gear. In fact, but we were taken to the baths to get pit hats and boots. Perhaps it was something to do with a song one of them called Sixteen Tons.

It was a real touch of luxury to be in pit overalls. Back at the barn at Berry Hill I'd seen some miners wearing ragged trousers with their waistcoats hanging out.

The lads at Kemball were all fighting to be proud of the boots. The majority got through the training, but some dropped out because of the time they had to get up in the morning to get the pit bus.

At that time, bus will wait down the pit because it ran in the fields, close when their pit tops come to stay there. At Berry Hill I did a job in the last pit like pit where any better paid.

But in the 1950s, if you learned a trade you might would be about £5 or £5.95. In wages in a pit it would be more than That was the big attraction.

The prospects were brilliant even if you'd education and a high. You could go to college eventually, if you wanted to the coal working life in three of you finally seemed to planned.

At Kemball, which was then known as Stoke's Pit, trainees were started with machinery work, the industry very soon had to be towards the coal face working life what they did and their efforts to do anything else.

We just to do the dirt to our first lesson, which meant to tap a tub to unhitching it facing a train that the guides of the wheel it...

lock them in their way side didn't leave holes.

We did this all day long and after a few days just weren't there any more or a time. But you had to be careful to keep your fingers and thumbs out of the way, or else I've ended on the backside of the tub.

To set them quick I put a backward finger this way. So I was also the medical centre to check they had never had a broken bone, which he said to make the floor out of a couple inches.

There was a lot of safety training at Kemball. They were shown the importance of clothes properly, like working through coal dust. The head snapped here of four inches only. You got it jammed, but it had to be seen, or your finger kept on working.

Kemball put us were taught how to stay into seeing quarry work as the looking of the coal surface area. If a box which with chains in our hours and to research the tubs in a job proper way chained with any levers up.

Through many of us have been over as by the dollars to we had up to the pit.

$$\text{THERE}$$ was a lot on safety. They was first and we didn't know these practical things,

using like that we're years to have as open on their edges.

Some of then felt skinny. They know it all when it came along of learning. But the "boys" who were told belong to that by their older ones, most of whom went to the plus.

On the pit itself they were impressed. We all have been given a tally with our number on their from you if he was on with the hooks down. There was a list of the both fifties and appeal on the tops. Making sure on what and then was one pair.

Anyone wanting to smoke they were to be the chance to go on the chance which was the gate.

In an old I was there up to Berry left the mines were matched in most-time we all saw around the mines. It was a kind of monthly belong Kemball in Sentinel.

The Mine Workers were 150 known at Kemball at it a Notable before it was to become the news hostel...

with came the miner hostel but some were lived years too...

...don't

We're the Kemball 1947 we're very pressed as how new look mines...

8

Lads so proud to be in pit gear

As soon as we arrived at Kemball for training, we were given the basics – pit hat, pit boots with steel toecaps and a pair of bib and brace overalls.

We were really proud of wearing that gear. So proud, in fact, that we went on the bus in our pit hats and boots. Perhaps it was something to do with a song at that time called Sixteen Tons.

It was a real touch of luxury for us to get overalls. Back at my base at Berry Hill, I'd seen older miners wearing ragged trousers with their backsides hanging out!

The lads at Kemball were all fighting to be cock of the roost. The majority got through the training, but some dropped out because of the time you had to get up in the morning to get the pit bus.

At that time, lads still went down the pit because it ran in the family, even when their parents tried to stop them. At Berry Hill I did a job in the part of the pit where my father worked.

But in the 1950s, if you learned a trade your wages would be about £2 10s (£2.50) whereas in the pit it would be double that. This was the No 1 attraction.

The prospects were brilliant, even if your education wasn't right. You could go to college immediately if you wanted to get on. Your working life in front of you already seemed to be planned.

At Kemball, which was once known as Duke's Pits, trainees always started with haulage work. Incidentally, men often went back to it towards the end of their working life when they were too worn out to do anything else!

We went up the dirt tip for our first lesson, which was how to stop a tub by lockering or throwing a metal bar into the spokes of the wheel to lock them. In those days tubs didn't have brakes.

We did this all day long and after a few days you could throw two bars at a time. But you had to be careful to keep your fingers and thumbs out of the way, or they'd be crushed on the underside of the tubs.

In my first week I got a blackened finger this way. So it was into the medical centre. In those days the sister had a Bunsen burner, which she used to make the blunt end of a needle red-hot.

You put your injured finger on a slab and, with the help of pliers, the nurse pushed the needle through your nail. The blood spurted three or four inches into the air. It was painful, but it had to be done, or your finger kept on throbbing.

Another job we were taught – and a highly-skilled one – was lashing on and lashing off, which involved using a 10-foot chain to attach the tubs to a moving rope while going up or down a dip.

There was a lot on safety, health and first aid too. But besides these practical things, young lads also had to learn to have respect for their elders.

Some of these lads thought they knew it all after a short period of training. But the "facey" ones were soon brought to heel by their older work-mates in a way which always worked.

Lads so proud to be in pit gear

At the pit bottom there were drums full of dirty black grease used on tubs wheels and points. If a lad had too much cheek, it was down with his trousers and the grease was daubed all over his lower parts.

Another method of dealing with a cheeky lad was to hang him up by his donkey coat on a dog nail.

I have mentioned that we had lessons at Kemball on pit safety, which reminds me of a real panic at Berry Hill when odd cigarette ends and matches were found underground.

For obvious reasons, there was a total ban on taking cigarettes and matches down below. You were searched at the lamphouse and again at the cage. Anybody caught with them was instantly dismissed.

So after this discovery at Berry Hill we were triple searched for weeks. We all suspected each other. It was a case of everybody being treated as guilty until proven innocent.

Then the mystery was solved. After having a last smoke before going down the pit, someone had thrown cigarette ends and used matches into a tub on the surface. Later, when the tub was underground they had fallen out onto the roadway.

So in the end nobody had been guilty of putting other men's lives at risk. But the episode made an impression on everybody and reminded us young lads that there were some rules which couldn't be broken.

The Way We Were

Pitheads and smoking chimneys of Florence Colliery in pre-war times.

Riding a runaway underground train

RUNAWAY trains were among the everyday hazards of working life when **GEORGE SHUFFLEBOTHAM** was a senior overman at Florence Colliery.

In this latest article, George remembers his own experience on an underground wagon train, which ran out of control, along with other incidents in the pit named after the daughter of the Duke of Sutherland.

The link-up of Florence and Hem Heath collieries, half-a-mile underground, on May 23, 1979. (Picture from Mining Memories by Fred Leigh)

'The boss wouldn't accept any of the explanations put forward. "I'll tell you the reason," he shrieked. "It's because these men are not terrified of the senior overman!"'

More Mining Memories next month

9

Riding a runaway underground train

Sitting in my old rocking-chair, I walk for miles and miles in my head, going along all those old underground roadways.

At Florence Colliery, there must be millions of pounds worth of equipment still lying down below – engines, manriders, haulages, the bullet train all smashed up. I'll come back to that later.

First, let me tell you about my own close shave on a runaway train. A big rope haulage had just been installed on a one-in-four roadway, which is pretty steep. I had to try it out loaded with materials.

I signalled with the bellwire to go down with several tons of steel on the back. The train was designed to travel at two miles an hour and I had to go half-a mile down the dip.

All of a sudden, it started to pick up speed – 4mph, 6mph, then 15mph. I realised that I was riding on a runaway. The sides of the roadway became a blur. I tried swinging on the signal wire overhead, but only burnt my hands.

I knew I had to get off and bailed out in the confined space, jumping straight into a refuge hole. The cars went on down the dip, hit a derrick and turned upside down. Everything was smashed up, but luckily it didn't reach the men at the bottom.

The management said I must have done something wrong, as this couldn't happen by itself. So they did a reconstruction of the incident, minus me, of course. They were astonished when the same thing happened again. The defect was traced to the rope wrapped round the drum.

Some years later, a lad did lose his life on this same haulage car in different circumstances.

Now that bullet train, which was so called because it travelled so quickly by haulage standards, about 5-6 mph. This was a manriding haulage to Trentham Pit in a roadway which was extremely hot.

As a temporary measure, we had just put up regulating doors for ventilation halfway down the dip. But the two lads on the bullet train forgot the doors were there and the trolleys went straight through them, turning them into splinters.

Perhaps the lads hadn't been paying attention, but the boss wouldn't accept any of the explanations put forward.

"I'll tell you the reason," he shrieked. "It's because these men are not terrified of the senior overman!"

Here's a tale which I'll call an act of God. A loco driver was bringing in a large electric motor to replace one on a coal face. He stopped by a telephone to wait for a call from us to proceed.

However, he didn't know about an emergency at the other end. A trolley carrying a gearbox weighing several tons broke away while being shunted. Travelling at high speed, it smashed through two sets of doors as it ran down the track towards the standing loco.

The loco driver, still unaware of what was going on, got out of his cab to use the telephone, which was in a refuge hole. As he picked up the phone, the runaway trolley hit the loco.

The gearbox went through the one-inch plate, landing on the driver's seat. He had been sitting there only a few seconds before and would have been killed outright. Yes, it was an act of God all right.

Riding a runaway underground train

You may not know that Florence Colliery dated back to the 1870s, which reminds me of some old workings which ceased production soon after the Second World War.

This district was kept open as a second means of egress, or exit, in an emergency. All the old tubs, rail works, winding and haulage engines were still in place. You could wander round for miles.

By some old loading points there were a lot of heavy wooden beams and carved on them were the dates of land and sea battles which took place during the war. One carving, I remember, recorded the sinking of HMS Hood, but there were numerous others.

They were very probably done by the Bevin Boys who worked at Florence, and other local pits, during the war. It was their record of what was happening out in the battlefields.

The Sentinel, Saturday, July 26, 1997

The Way We Were

I could have gone down the pit at six

AS I waited to go down the pitshaft at Berry Hill, I often thought back to the time when I was about six and my father worked a double shift in the pit bottom.

I went to the pit on a Sunday with my elder sister Beryl to take my father's dinner, kept warm between two tin plates wrapped with string, and a bottle of cold tea.

As the banksman knocked for the cage to come up the shaft, we would be standing just a few feet from it. The speed and ferocity of the cage coming to the surface terrified me.

My father's dinner was put in the cage and on one occasion I even was going down at the same time. As a treat, the banksman said I could go down the shaft to see my dad, but I declined the offer. I can't imagine anything like that happening today?

We didn't have water or electricity at home, so I had to fetch a bucket of water from the pit top at about 7.30am every day. One particular morning around 1949 I rounded the pit bank and couldn't believe my eyes. The whole of Top Pit Colliery was on fire.

Fire engines came from Hanley and Longton, but took a long time to get there along cart tracks with low bridges. I remember being upset when the foremen brought down nests of young birds and eggs from the eaves of the burning engine house.

SO I was brought up breathing pit air and it was inevitable that I should follow in my father's footsteps and start work at Berry Hill Colliery at 15.

As a matter of fact, the banksman I knew as a child actually sent me down the pit to work. I'll never forget that character with his oily cloth cap, thick black belt round his waist and trousers tied below the knee with string to stop them going up his legs.

At 18 I was a fully-qualified faceman. One night the coal cutter's mate didn't turn up, so I worked with Shiller Barker, who had a reputation for working till a job was done, sometimes for 24 hours at a stretch.

If he had a breakdown on the face, he would work for evermore and then nod off to sleep on his way home to Kingsley on the bus.

Once somebody was climbing the stairs on the bus and disturbed him as he slept. Shiller jumped up, accidentally knocking the man down the stairs. "Sorry, mate," he said. "I thought the roof was coming in!"

As I said, I was working with Shiller on the coal cutter one night when the picks on the cutter struck hard rock underneath the coal seam.

It was the Cockshead seam and there was a bit of gas, which was ignited by sparks from the cutter. Flames shot out towards me and then flashed 100 yards down the face before burning out. It was like a ball of fire, or a Will o' the Wisp.

AT Berry Hill we had a fair but very tough manager named Frank Stevenson. It was no use trying to threaten this man. If you had an argument with him in his office, he just said he would settle it outside

●The waste tip at Mossfield Colliery

with you. Either that or pick your window.

Although there were some hard men at Berry Hill, they didn't get into a confrontation with Stevenson. But he was a fair man and he went on to become NCB area manager.

The next manager was Reg Beaton, and he soon made it clear that we'd be a new brush sweeping clean. He cut rates and jobs. Perhaps he could see that the closure of Berry Hill was coming.

So in 1958 or 1959 the union men called a mass meeting and hired the courtroom at Hanley Town Hall. Everybody wanted Beaton to be shifted, but nothing came of it.

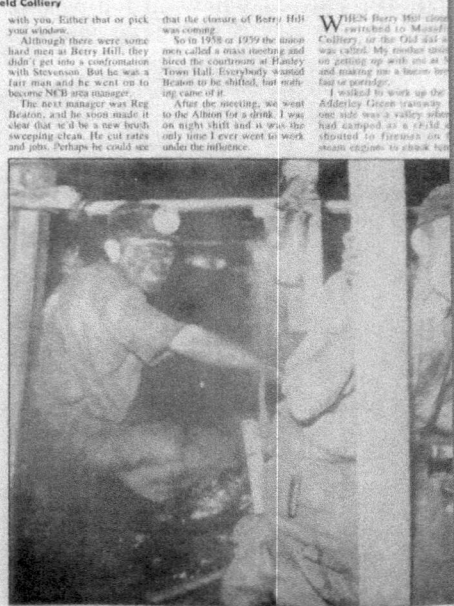

After the meeting, we went to the Albion for a drink. I was on night shift and it was the only time I ever went to work under the influence.

WHEN Berry Hill closed I switched to Mossfield Colliery, or the Old stax as was called. My mother stuck on getting up with me at 5 and making me a bacon breakfast or porridge.

I walked to work up the Adderley Green tramway, one side was a valley where had camped as a child and shouted to firemen on the steam engines to chuck lumps

> 'Once somebody was climbing the stairs on the bus and disturbed him as he slept. Shiller jumped up, accidentally knocking the man down the stairs. "Sorry, mate," he said. "I thought the roof was coming in!" '

●Two miners use an Anderson Boyes electric coal cutter of the 1930s. With it, the length during a seven-hour shift. Pictures from Mining Memories, A Portrait Staffordshire by Fred Leigh (S.B. Publications)

10

I could have gone down the pit at six

As I waited to go down the pitshaft at Berry Hill, I often thought back to the time when I was about six and my father worked a double shift in the pit bottom.

I went to the pit on a Sunday with my elder sister Beryl to take my father's dinner, kept warm between two tin plates wrapped with string, and a bottle of cold tea.

As the banksman knocked for the cage to come up the shaft, we would be standing just a few feet from it. The speed and ferocity of the cage coming to the surface terrified me.

My father's dinner was put in the cage and on one occasion a man was going down at the same time. As a treat, the banksman said I could go down the shaft to see my dad, but I declined the offer. I can't imagine anything like that happening today.

We didn't have water or electricity at home, so I had to fetch a bucket of water from the pit top at about 7.30am every day. One particular morning around 1949 I rounded the pitbank and couldn't believe my eyes. The whole of Top Pit Colliery was on fire.

Fire engines came from Hanley and Longton, but took a long time to get there along cart tracks with low bridges. I remember being upset when the firemen brought down nests of young birds and eggs from the eaves of the burning engine house.

PIT LIFE: *I would do it all over again ...*

Coal Cutting of Yesterday
An Anderson Boyes electric coal cutter of the 1930s. It was capable of undercutting a 150-yard length in a 7-hour shift with a team of two men.

So I was brought up breathing pit air and it was inevitable that I should follow in my father's footsteps and start work at Berry Hill Colliery at 15.

As a matter of fact, the banksman I knew as a child actually sent me down the pit to work. I'll never forget that character with his oily cloth cap, thick black belt around his waist and trousers tied below the knee with string to stop dust going up his legs.

At 18 I was a fully-qualified faceman. One night the coal cutter's mate didn't turn up, so I worked with Shiller Barker who had a reputation for working till a job was done, sometimes for 24 hours a stretch.

If he had a breakdown on the face, he would work for

evermore and then nod off to sleep on his way home to Kingsley on the bus.

Once somebody was climbing the stairs on the bus and disturbed him as he slept. Shiller jumped up, accidentally knocking the man down the stairs, "Sorry mate," he said. "I thought the roof was coming in!"

As I said, I was working with Shiller on the coal cutter one night when the picks on the cutter struck hard rock underneath the coal seam.

It was the Cockshead seam and there was a bit of gas, which was ignited by sparks from the cutter. Flames shot out towards me and then flashed 100 yards down the face before burning out. It was like a ball of fire, or a Will o' the Wisp.

At Berry Hill we had a fair but very tough manager named Frank Stevenson. It was no use trying to threaten this man. If you had an argument with him in his office, he just said he would settle it outside with you. Either that or pick your window.

Although there were some hard men at Berry Hill, they didn't get into a confrontation with Stevenson. But he was a fair man and he went on to become NCB area manager.

The next manager was Reg Beaton, and he soon made it clear that he'd be a new brush sweeping clean. He cut rates and jobs. Perhaps he could see that the closure of Berry Hill was coming.

So in 1958 or 1959 the union men called a mass meeting and hired the courtroom at Hanley Town Hall. Everybody wanted Beaton to be shifted, but nothing came of it.

After the meeting, we went to the Albion for a drink. I was on night shift and it was the only time I ever went to work under the influence.

When Berry Hill closed I switched to Mossfield Colliery, or

PIT LIFE: *I would dc it all over again ...*

the Old Sal as it was called. My mother insisted on getting up with me at 5am and making me bacon breakfast or porridge.

I walked to work up the old Adderley Green tramway. On one side was a valley where I had camped as a child and shouted to fireman on big steam engines to chuck lumps of coal off the tender for our fire.

At Mossfield in the early 1960s, I used to see umpteen people picking coal on the tip. They had prams, bikes and wooden trucks made from orange boxes. But coal-picking could be a dangerous business.

The top of the tip was cone-shaped and about 200 feet high. A cage went up a track to the top with two or three wagons of waste. Big lumps went careering down the tip and coal pickers were often injured.

When I got home I sometimes had to get the coal in myself. Once a month a pile of coal was tipped on the ground about 80 yards from our house. I had to collect the lot in buckets and a tin bath.

The Way We Were

Spike, Tummy and Herby
... last of the funny kind

Miners enjoying leisure time in the fresh air after working in the dusty atmosphere of a pit
Picture from The Miners by Anthony Burton (Futura Publications Ltd)

Berry Hill Colliery in its final days before demolition in the 1960s.
Picture from Fred Leigh's book 'Mining Memories'

More Mining Memories next month

11

Spike, Tummy and Herby ... last of the funny kind

Like most pits, Berry Hill had a whippet club which ran races in a field by the old and long-closed Brookhouse Colliery.

The course went straight across the field and the traps were kept there permanently. Bets were made on every race. The dogs all had names, which were probably changed regularly, so nobody knew if there was a good dog taking part.

It eventually died out and with it went the characters. I'm thinking particularly of three Berry Hill men I knew called Spike, Tummy and Herby, who were like those three in the Last of the Summer Wine.

It was early 1950s and I'd just started at the pit when I met these three from Bucknall, whose hobbies seemed to consist mainly of whippeting, rabbiting and boozing.

They could be seen on most days walking round the countryside at Berry Hill, each in charge of one or two whippets and with their cloth caps hanging off the side of their heads.

The farmers would be tearing their hair, but it made no difference to these three, who were after rabbits for the pot. The farmers just had to grin and bear it.

When Spike, Tummy and Herby did feel like going to work, they mostly turned up for the night shift. If you heard that they were coming into your district, you knew you'd have a full night's entertainment.

One particular story about Tummy went the rounds at

PIT LIFE: *I would do it all over again ...*

Berry Hill. After a dinner-time booze-up, Tummy staggered home and was met by his missus, who hit him on top of the head with a cast-iron saucepan.

Tummy should really have gone to hospital, but he stayed away because he saw there was a chance to make some money. He managed with padding on his head until it was time for him to go to work. He perched his cap on the lump and lacerations and went off to the pit. He went underground with his pit hat balanced precariously on top of his head.

Berry Hill Colliery.
William Jack Collection

Only a few minutes later, the word got round that Tummy had been injured by something which had struck him on the head. He was rushed out of the pit and up to the hospital.

Tummy must have been smiling when the Coal Board paid him quite a few bob in compensation for an injury sustained at work!

There's also a comical story about Spike which I believe is true. He was spotted walking in Bucknall with a duck, which he was holding with a piece of string round the neck.

When a policeman asked him if he had pinched the duck, he replied that he was merely taking it for a walk. Then he added: "You can't blame me for all this lot" and pointed to eight ducklings which were following close behind!

Talking of people sailing close to the wind reminds me of Arthur, one of the bosses at Wolstanton Colliery. He used to come to the pit in an old van and often departed with it full of contraband, loaded up by somebody else to his order.

One night we had a major roof fall which closed one end of the face. For once, Arthur had to work overtime and his luck ran out. His van, loaded with coal and other stuff, was parked in the manager's garage. The manager arrived early and found it.

Everybody thought Arthur's day of reckoning had come. But he was a crafty customer who knew a lot of people's secrets. After a word with the manager, he emerged with his job still intact.

I should add, by the way that Arthur was a Robin Hood type of character who often took coal away for hard-up pensioners he knew. He rarely took things for himself.

Going back to knocks on the head, I must tell you about an incident when we were driving a tunnel from Florence Colliery to Hem Heath.

I had deployed two men, Ron and his mate, to make a pumping station halfway down a one-in-four dip. They had to knock in some reinforced girders to support the roof.

Ron, a big strapping youth, held the girder steady on his head while his mate hit the end of it with a seven-pound hammer. Ron took a fresh hold, lifting the girder and moving slightly.

The hammer came crashing down again and this time it struck Ron right between the eyes. I ran to him as he was on his knees. He said: "I've never seen colours like this in all my life" and then passed out.

Ron had a lump on his face as big as a cup. We rushed him up to the hospital and he recovered. It reminded me of the man who said: "When I nod my head, hit it".

The Way We Were

CONCLUDING OUR POPULAR SERIES . . . MINING MEMORIES

Pit life? I would do it all over again . . .

LIKE thousands of lads in North Staffordshire, I followed my father into the pit and if I could have my life over again I'd do the same.

Most of all, I enjoyed the comradeship of my workmates. You could have a damn good row in the pit and the next minute you were mates again.

For most of us it was a way of life, not a job. It was how my family had been brought up and when the time came I just took it on. I worked at Berry Hill, Mossfield, Wolstanton, Hem Heath and Florence and I don't regret any of it – I'd do it all over again.

Mind you, it was a very hazardous job. There were so many working together in confined spaces. We had to look after each other. Hardly a day went by without somebody getting hurt in some way, whether it was a busted finger or a stretcher case.

The pit officials were very strict in my early days. The senior overmen were old miners who'd been there in the pit since the 1920s. They kept everybody in line. They were almost like gods.

To get a regular job on coal you had to prove yourself. Once you'd proved yourself in a team of 20 men you'd got to prove yourself to the butty (chargehand). And you'd certainly got to show the senior overman you could do the job.

It was damned hard work for a young lad. However, in those days the pubs shut at 10 o'clock. There were no night clubs. We were in bed at half-past ten. Otherwise, there was no way we could have got up at five in the morning and chucked a buttock of coal on.

In my first few weeks at Berry Hill I learned to chew tobacco and take snuff. I never stopped chewing tobacco till the day I finished in the mines. It makes you breathe through your nose and the snuff keeps your nostrils clear.

They say that in the pit you either lose your teeth or lose your lungs. I've proved that to be true myself.

I STARTED in the 1950s in a period of change and modernisation. At Berry Hill we were working in the pick and shovel era and using old methods of conveying coal out of the pit. Everything was hand-loaded. There was no machinery except a coal cutter.

Then a few years later we went off the butty system and on to a national power-loading agreement where everybody got the same, no matter how hard you worked.

All the same, we weren't well off. I couldn't afford a car,

or even a motorbike. In the 1950s you could have counted on your fingers the number of cars outside any local colliery.

Public transport was still booming. Most bus stops in the Potteries had a queue of miners waiting for designated pit buses, which were paid for by the Coal Board.

You'd see them in cloth caps, snappin' wrapped in Sentinel paper under arms or squeezed into jacket pockets. There was the usual banter and a few choice words. But on the rare occasions that a woman was at the bus stop there wouldn't be a peep out of them.

When you finally got onto a crowded bus you couldn't use for tobacco smoke. In those days most people smoked. Since the transport was free, you couldn't really complain.

Hard work underground made us very hungry. We thought nothing of taking eight to ten rounds of bread to work, or half a loaf. Then after an early shift we could have a roast dinner for a shilling, or meat and potato pie with gravy for tuppence or threepence.

In those days meals were supplied at cost and as I've said, transport was free. The Coal Board cared about their workers then. We seemed to lose that along the way

IN CONCLUSION I want to say that until I was asked to do these stories 12 months ago I had almost forgotten about my past working life. But the stories have come back and many people have rung me to say that they remember them, too. It's been very worthwhile.

The history of mining in North Staffordshire deserved to be recorded. There are so many tales to be told. I'm just one man who has thought of a few stories that I remember from my 40 odd years in the pit.

But there are literally thousands more.

● **North Staffordshire mining veteran William Clay, who worked underground for over 60 years.**

●Day of triumph at Florence Colliery, George Shufflebotham's last pit.

Marking the site of Hanley Deep Pit.

12

Pit Life? I would do it all over again ...

Like thousands of lads in North Staffordshire, I followed my father into the pit and if I could have my life over again I'd do the same.

Most of all I enjoyed the comradeship of my workmates. You could have a damn good row in the pit and the next minute you were mates again.

For most of us it was a way of life, not a job. It was how my family had been brought up and when the time came I just took it on. I worked at Berry Hill, Mossfield, Wolstanton, Hem Heath and Florence and I don't regret any of it. I'd do it all over again.

Mind you, it was a very hazardous job. There were so many working together in confined spaces. We had to look after each other. Hardly a day went by without somebody getting hurt in some way, whether it was a bosted finger or a stretcher case.

The pit officials were very strict in my early days. The senior overmen were old miners who'd been there in the pit since the 1920's. They kept everybody in line. They were almost like gods.

To get a regular job on coal you had to prove yourself. Once you'd proved yourself in a team of 20 men you'd got to prove yourself to the butty (chargehand). And you'd certainly got to show the senior overman you could do the job.

PIT LIFE: *I would do it all over again ...*

It was dammed hard work for a young lad. However, in those days the pubs shut at 10 o'clock. There were no night clubs. We were in bed at half past ten. Otherwise, there was no way we could have got up at five in the morning and chucked a buttock of coal on.

In my first few weeks at Berry Hill I learned to chew tobacco and take snuff. I never stopped chewing tobacco till the day I finished in the mines. It makes you breathe through your nose and snuff keeps your nostrils clear. They say that in the pit you either lose your teeth or loose your lungs. I've proved that to be true myself.

I started in the 1950's in a period of change and modernisation. At Berry Hill we were working in the pick and shovel era and using old methods of conveying coal out of the pit. Everything was hand loaded. There was no machinery except a coal cutter.

Then a few years later we went off the butty system and on to a national power-loading agreement where everybody got the same, no matter how hard you worked.

Pit Life? I would do it all over again.
Courtesy of *The Sentinel*.

Pit Life? I would do it all over again ...

All the same, we weren't well off. I couldn't afford a car, or even a motorbike. In the 1950's you could have counted on your fingers the number of cars outside any local colliery.

Public transport was still booming. Most bus stops in the Potteries had a queue of miners waiting for designated pit buses, which were paid for by the coal board.

You'd see them in cloth caps, snappin' wrapped in the Sentinel paper under arms or squeezed into pockets. There was the usual banter and a few choice words. But on the rare occasions that a woman was at the bus stop there wouldn't be a peep out of them.

When you finally got into a crowded bus you couldn't see for tobacco smoke. In those days most people smoked. Since the transport was free, you couldn't really complain.

Hard work underground made us very hungry. We thought nothing of taking eight to ten rounds of bread to work, or half a loaf. Then after an early shift we could have a roast dinner for a shilling, or meat and potato pie with gravy for twopence or threepence.

In those days meals were supplied at cost and as I've said, transport was free. The Coal Board cared about their workers then. We seemed to lose that along the way.

In conclusion, I want to say that until I was asked to do these stories 12 months ago I had almost forgotten about my past working life. But the stories have come back and many people have rung me to say they remember them, too. It's been very worthwhile.

The history of Mining in North Staffordshire deserved to be recorded. There are so many tales to be told. I'm just one man who has thought of a few stories that I remember from my 40-odd years in the pit.

But there are literally thousands more to be told.

PIT LIFE: *I would do it all over again ...*

Florence Miners.

Pit Life? I would do it all over again ...

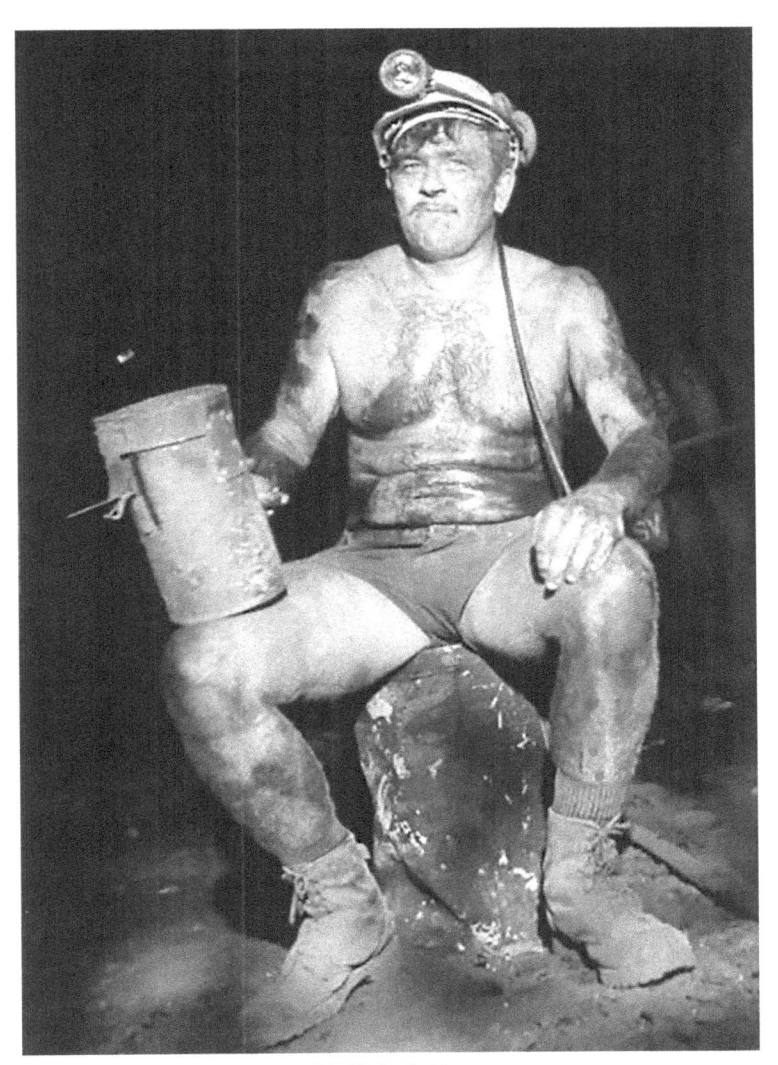

Mr Kevin Oakley.

All your yesterday

You and your

Passage to Florence

PIT BANNER: Miners pictured at the breakthrough point in 1979 when Florence Colliery and Trentham Pit were linked via an underground tunnel.

Traffic thundered down High Street

TRAFFIC PROBLEM: Vehicles nose-to-tail were a common sight in Newcastle High Street when there was no alternative route for through traffic. This picture was taken around 1960.

13

Passage to Florence

Late addition 6 years later

Coal mining still seemed to have a long term future in North Staffordshire when two major collieries, Florence and Trentham were joined up underground in 1979. At that time well over 3,000 people were employed at the two high production pits, which were linked half a mile below ground by twin tunnels enabling the entire output of coal at Florence to be transported to the surface via Trentham.

George Shufflebotham, then a colliery deputy at Florence, was in daily charge of teams working on the tunnelling project, which took 18 months and sometimes progressed only about three meters in 24 hours.

"We drove the tunnels for over half a mile from our end and the job was difficult, hot and wet", he says.

"We were drilling and firing down a one in four drift and had problems with different seams, soft and hard ground and a great deal of water."

"When the job was finished it made great savings. At that time no one thought that either of these big pits would close. At Florence we had a coal face which was nearly under Stone and there was talk of extending it to Cannock Chase."

George, who later became senior colliery overman, recalls that Florence's output topped one million tonnes for the year

PIT LIFE: *I would do it all over again ...*

1987/88, yet the pit was closed five years later and Trentham followed soon afterwards.

He says, "The equipment left down there must have been worth millions of pounds."

All our Yesterday's with John Abberley.

George Shufflebotham's lamps and keys ...

Passage to Florence

Boys from no3 Florence pit outside Hem Heath drift mine following a fundraiser for Kemball Special School, 1991.
Left to right: *Mick Salih, Kevin Oakley, Kenny Burgess, Wayne Rowley and Steve Rose*

Boys from no3 pit underground.
Left to right: *Mick Salih, Kevin Oakley, Kenny Burgess, Wayne Rowley and Steve Rose*

PIT LIFE: *I would do it all over again ...*

Florence.

Mr Keven Oakley.

Lads so proud to be in pit gear

Riding the runaway underground train.

Florence bowling alley.

Lightning Source UK Ltd.
Milton Keynes UK
UKHW021120230821
389329UK00015B/1273